STEAM AT WORK
Road and Farm Engines

Sentinel tractor converted from 6/7-ton wagon No 3237 built c1927, working for White, Tomkins & Courage Ltd, millers, showing a full head of steam in Liverpool, May 1947

STEAM AT WORK
Road and Farm Engines

Anthony Beaumont

DAVID & CHARLES
Newton Abbot   London   North Pomfret (Vt)

By the same author
Traction Engine Pictures (4th imp)
Traction Engines on Parade
Rally Traction Engines (2nd imp)
Traction Engine Prints
Steam Up! Engine and Wagon Pictures
The Organs and Engines of Thursford
Fair Organs (2nd imp)
A Gallery of Old Timers (2nd imp)
Traction Engines and Steam Vehicles in Pictures
Ransomes Steam Engines; An Illustrated History
Fairground Steam
Traction Engines Past and Present
Steam on the Road

British Library Cataloguing in Publication Data

Beaumont, Anthony
  Steam at work.
  1. Traction-engines – History – Great Britain –
  Pictorial works
  I. Title
  629.2'292'0941   TJ700

  ISBN 0-7153-8121-0

  Library of Congress Catalog Card Number 81-67002

Phototypeset by Typesetters (Birmingham) Limited
and printed in Great Britain
by Biddles Limited, Guildford
for David & Charles (Publishers) Limited
Brunel House Newton Abbot Devon

Published in the United States of America
by David & Charles Inc
North Pomfret Vermont 05053 USA

CONTENTS

INTRODUCTION   6

TRACTION ENGINES   7

STEAM TRACTORS   26

PLOUGHING ENGINES   36

SHOWMANS ROAD LOCOMOTIVES   47

STEAM WAGONS   57

STEAM ROLLERS   76

INDEX TO PLATES   95

ABBREVIATIONS

SC single cylinder.
DCC double-crank compound cylinders.
PE ploughing engine.
psi pounds per square inch.
nhp nominal horsepower. This originates from the Royal Agricultural Society of England
when steam power began to replace horses. It was an attempt to equate steam engines with
work done by horses but is an arbitrary figure based only on the steam piston area. Brake
horsepower at the flywheel is approximately five to eight times the nhp figure and maximum
for short emergency power bursts.

COPYRIGHT

The copyright of all photographs originating since 1931 is vested in the owners of the
negatives from which prints for this book have been specially made.

ACKNOWLEDGEMENTS FOR PHOTOGRAPHS

John P. Mullett for the greater part of the pictures.
J. L. Thomas for two Sentinel pictures on page 71.
Fodens Ltd for two 'undertype' pictures on pages 67 and 68.
The remaining photographs are from the author's collection.

INTRODUCTION

Steam road transport, haulage and agricultural work had greatly declined by the mid-1930s although its demise was long drawn out. The longevity of the steam engine and the high standards of its builders enabled owners to continue with obsolete but reliable prime movers which were often well-liked.

Inevitably the time approaches when the actual working régime of the traction engine family will have passed from living memory. Certainly the steam preservation movement has helped in several ways in explaining the working functions of various types of traction engines. Unfortunately the scenes of earlier times cannot be recaptured except in the comparatively small number of technically good photographs.

All the pictures except one or two are here published for the first time. Included are several unique prints made directly from a number of 8½in × 6½in glass negatives c1907. These were recently saved by chance during the demolition of an engineering works.

During the decline of steam it was the outstanding designs of steam wagons and rollers which lasted longest in commercial use. The final types of both were exported in 1950; the rollers by Vickers Armstrong Ltd, Newcastle. Sentinel steam wagons from that company at Shrewsbury went to the Argentine.

Many county councils continued with steam rollers to the mid-1950s if only because the engines were so well-maintained by the same drivers during many years.

Amusement caterers, faced with rapidly rising labour and maintenance costs after 1945, changed to the diesel lorry/generator, leaving some fine showmans road locomotives to the fate of the cutting torch. The last commercial use of a showmans engine was in 1957 when a Foster owned by Pat Collins appeared at the Birmingham Onion Fair.

The humbler but versatile traction engine (unfortunately a present journalistic name for every steam engine with road wheels and a chimney) found sporadic farm use during the 1939–45 War. Obviously steam could not compete economically with internal-combustion tractors having power take-off pulleys. In any case, combine harvesting was then revolutionising the traditional thrashing from the harvested stack.

The majority of traction engine builders had ceased producing them by 1934. Fosters of Lincoln made the last production traction in 1942.

The double-engine cable system of steam ploughing and varieties of cultivation introduced by John Fowler in 1864 remained in widespread use, unchanged in basic design, until 1960. At that date, on the death of John Patten—the last significant steam ploughing contractor, of Little Hadham, Hertfordshire —all his engines and tackle were sold piecemeal.

Despite the physically demanding and grimy work of employed steam men in an era of low wages, their expressions before the camera were seldom dour. The appearance of the men and engines in their working years, even in adverse conditions, seems evidence enough of essential jobs being very well done.

Anthony Beaumont
South Wootton
June 1980

TRACTION ENGINES

The appearance of the general-purpose traction engine remains familiar to the public because of the long-standing steam preservation movement. The pictures of sixteen makes range from a unique 1859 photograph to the final designs. These changed little from the 1880s until production ceased in 1942.

Many makers classed their engines as 'agricultural', implying general farm use (excluding ploughing) and short haulage road work. Consequently they were low-geared, usually two-speed, unsprung and capable of averaging about 5mph on the level hauling loads of twenty tons. The 'speed bursts' of today were not a feature of working engines.

All these engine types, excluding some of the earliest mid-nineteenth-century designs, were based on a boiler of 'locomotive' design as used on the railways throughout their steam era. A simplified description of some of the principal constructional features adds interest to the pictures and supports additional details in the captions.

The cylindrical boiler barrel, made in sections, has a hollow extension smokebox front and a round- or flat-topped outer firebox at the rear. An inner box shape contains the fire itself, the heat from which passes to the boiler water through the surrounding water spaces and through about thirty-five tubes passing horizontally to the smokebox. All the boiler components were riveted (only welding is now permissible in a new boiler) but the tubes are expanded because their replacement was almost a routine job. It can be done single-handed by a robust man using one special expander tool and the minimum of hand equipment.

All flat surfaces of the firebox plates are closely stayed together to resist boiler pressure. The firebox top is also strengthened by girder plates and stays, or in Marshall's design by diagonal corrugations.

The outer firebox plates are extended upwards and rearwards and braced across the boiler top and at the rear by the manstand and water tank below it. In total this makes a rigid box-form for the gear shafts and rear axle.

Single or compound (from 1882) cylinders are bolted to the reinforced boiler. The cylinder block base is either cast curved and hand-fitted or machined flat to suit a boiler seating of the same accuracy. The majority of makers arranged for the boiler steam to pass directly up and into the regulator chest via passageways in the cylinder casting. This usual design required a steam-tight joint of large area between boiler and cylinder base. A separate steam entry pipe was used by Ransomes.

A long connecting rod imparts a rotary motion to the crankshaft from the reciprocating piston via its rod, the crosshead and its flat or circular guides. This arrangement reduces wear because of the low value of the side thrusts.

Spur gearing from the crankshaft, through either one or two intermediate shafts, drives the rear axle. The rear wheels can be locked or free on the rear axle by means of manually operated large driving pins in the wheel centres.

Gear wheels on the first (crank) shaft and second shaft may be moved sideways on splines to engage fast or slow gear. There is also a position for free-running the motion for belt-driving work.

The engine should be stationary on level ground when gear changing. It can be done on the move by a skilled driver but clearly if a gear is 'missed' on a slope then the complete braking influence of the motion, together with counteracting steam on the piston if necessary, is lost. In these circumstances a 'runaway' engine may result. A rear-wheel rim brake lessens but does not avoid the risk.

Gear wheels are not used for reversing. This is done by reversing the movement of the piston in relation to the crankshaft. The steam-admission valve inside the flat covered valve chest is reciprocated by a crankshaft eccentric and its rod. A second identical eccentric set at a nearly opposite angle imparts the same but opposite motion to the valve when the driver uses the reversing lever to its full travel.

The mechanics of the eccentrics and their connections is the 'link motion' and in this form is the commonest of several types of reversing gear fitted to traction and allied engines.

Correct driving and steam economy includes using the reversing lever to varying but small extents to suit appropriate load conditions. These small movements of the reversing gear effect a reduced travel of the cylinder valve because its movement is then influenced by both eccentrics. Steam is then cut off at an earlier point in the stroke of the piston.

Stopping a traction engine without separate mechanical brakes is done in three ways. Closing the regulator normally stops the engine fast enough for most circumstances because all the motion work is geared up from the rear axle and therefore has a severe braking action. Closing the regulator and moving the reverse lever to 'mid-gear' adds braking power because the piston cannot complete its strokes without compressing air in the cylinder. Pulling the reverse lever quickly and fully over without closing the regulator immediately brings full steam pressure against the piston's movement. This is an emergency stop procedure and places severe loads on the motion work.

On the rear axle or the shaft next to it, a large bevel-geared differential was fitted to all the comparatively later engines. In common with motor vehicles, traction engines can therefore turn with both rear wheels driven at the correct and differing speeds according to the radius of the turn.

A large flanged drum on the rear axle carries a standard fifty yards of steel cable for haulage in severe conditions. In these cases the traction engine is stationary and the large driving pins are removed from the rear wheel centres. Cable haulage is used for extra-heavy wheeled loads in bogged-down conditions, timber winching, hauling and tree pulling.

Steam exerts force on a stationary piston, hence traction engine types can make a start from rest without the internal combustion engine's necessities of 'revving up', a clutch and a gearbox. This steam starting facility holds good even with heavy loads.

A single-cylinder double-acting steam engine makes two power strokes per revolution; compound cylinders make four overlapping power strokes. Should a 'single' traction engine stop with its crank on the forward or back 'dead centre', then the driver inches round the flywheel by hand so that a start can be made.

This is less likely to happen with compound cylinders but because steam enters the high-pressure (and smaller) cylinder first, the engine will not start should its high-pressure crank stop on either dead centre. Engine designers had a complete answer to this slight contretemps. A spring-loaded rod on the footplate operated a cylinder valve which momentarily passed steam direct to the low-pressure cylinder; its piston being at mid-stroke moved at once.

An important accessory when belt-driving is the cylinder-mounted governor gear. The governor itself is belt-driven from the crankshaft and is linked to a throttle valve to the steam entry into the cylinder block. The main regulator remains open when the governor is in action. Sudden increases in load cause the throttle valve to open wider before the flywheel speed can noticeably diminish. At different settings of the governor, therefore, steady flywheel speeds are maintained within the limit of the engine's capability to drive the machinery. Sawbench work and, to a lesser degree, thrashing, demonstrate the governor gear in action. Often the driver has left the footplate but the engine's exhaust note rises and falls automatically in a fascinating way.

Boiler fittings are obviously most important from the safety aspect. Above all, the safety valve mounted on the regulator chest cover must release steam at no more than the correct working pressure for the boiler. In early years 'lock up' safety valves were sometimes fitted to prevent unintelligent tampering in an effort to obtain more power. Nearly all later tractions were fitted with a double safety valve held down by a central spring and crossbar. One end of this bar is usually extended so that the valves may be released by hand (not when under steam!) in case they have stuck on their seatings.

Safe water levels in the boiler are shown by the upper and lower limits of the gauge glass fitted near the footplate and provided with three 'one-way' plug cocks. Two shut off the boiler exits and the third is for 'blowing down' the glass to remove any dirt or scum.

Should the water level become too low, the inner firebox top plate will overheat or 'burn' if the fire is going well. Prolonged overheating could cause a collapse of the firebox crown. To prevent any disaster in the event of gross mismanagement, the firebox crown is fitted with a screwed and detachable 'fusible plug'. This has a ⅜in diameter lead filling. Any overheating melts this lead and the contents of the boiler are violently ejected towards the fire. The engine and the immediate surroundings are hidden by steam.

Maintaining the boiler water level is done by the engine-driven pump and/or the steam-operated injector usually mounted near the manstand and rear wheel. The pump feeds only when the motion is working but the injector can be used at all times although it is normally in action when the motion is stationary. The rear water tank supplies both pump and injector and replenishment of the tank is arranged by a suction device. This will lift up to twelve feet from any convenient source. During thrashing, a youth was often hard-worked all day fetching buckets of water.

Allchin (Northampton) SC No 1285, 7nhp, four-shaft, built c1902, in steam at Hainsfort, Hertfordshire, in 1945 when owned by T. T. Boughton & Sons. The well-kept engine shows some paint lining, cross-arm governor and repair plate under the smokebox.

(*Above*) Aveling & Porter (Rochester) DCC road locomotive with top slide valves, probably 8nhp, hauling a Galloway stationary boiler. Details unknown but c1912 (the new-looking cycle is a Rudge-Whitworth with three-speed and acetylene lamp, all of that period). (*Below*) Brown & May (Devizes) showmans road locomotive DCC No 8742, 7nhp, built 1916. This sole-surviving showmans from this maker and with dynamo bracket cut off was used as a traction for thrashing until c1952. After standing it has been restored to showmans appearance and is here shown in 1944 when owned by J. H. Rundle of New Bolingbroke, Lincolnshire.

William Bray (Folkestone) 1857 traction shown in the Royal Arsenal, Woolwich, in 1859 and the only known photograph of this engine—his first of sixteen built up to 1861. Duplex cylinders drove both rear wheels which could be clutched independently. Bray's patent of protruding blades through the rear wheel rims was an eccentric and links arrangement; the blades protruded at any part of the circumference at will. Here they are fully out at the top and flush with the road at the bottom. Bray is the top-hatted figure at right.

Burrell (Thetford) SC No 1297, 7nhp, two-speed built 1887, still at work here thrashing at Coningsby, Lincolnshire, in September 1944 and owned by Smith of Dogdyke, Lincs. This fine old traction still had the original design of cross-arm governor and 'stove-pipe' chimney.

(*Above*) Clayton & Shuttleworth (Lincoln) DCC crane traction, convertible to roller using the forecarriage shown. The worm-geared crane hoist has lifted 4¼ tons for this c1905 builder's photograph and as indicated this and similar engines were for Chatham Dockyard. (*Below*) Cooper (King's Lynn) digger, 'Estate Class', first type, built 1900 and basically a DCC traction with a geared digging frame which was raised and lowered by a hydraulic cylinder. The forty prongs were chisel-shaped in the front row and curved in the back row. A 1900 Newcastle test showed 1¼ acres dug in 2½ hours.

Burrell single-crank compound No 2836, three-shaft, 7nhp, built 1906, thrashing from the stack and chaff cutting at Ford End Farm, Ivinghoe, Buckinghamshire, in 1948 when owned by A. T. Oliver. This is heavy work for tractions; a sharpened replacement chaff-cutting blade is leaning against the engine's rear wheel.

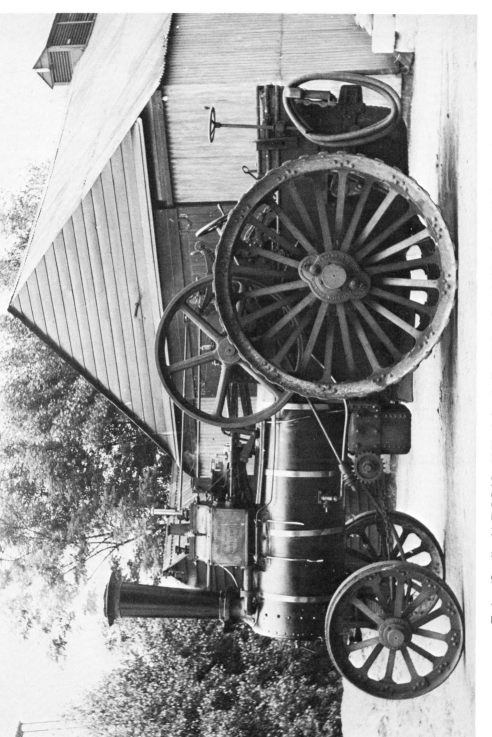

Foden (Sandbach) DCC No 9052, 8nhp, built 1919 and probably the last Foden traction made. The semi-elliptical springs behind the flywheel were the second type of springing. The small underboiler water tank and large rear wheels (twenty-four spokes here) are Foden characteristics. Shown c1945 when owned by C. W. Lambert of Horsmonden, Kent.

(*Above*) Foster (Lincoln) SC No 2827, built 1903, thrashing in September 1935 when owned by D. Stanton of Boston, Lincolnshire. The cross-arm governor is controlling the speed without the driver's presence and a chimney spark-arrestor is in place. Seen on the right are the sack scales and spare chaff-cutting knife. (*Below*) Fowler (Leeds) DCC No 9713 with top slide valves, probably 8nhp, built 1902. The small Fowler-type governor is belted up and the engine standing for belt-driving a rack sawbench using wood for fuel. Shown in 1944 and owned by B. Frank & Sharp Ltd, Ryedale Sawmills, Helmsley, Yorkshire.

Fowell (St Ives) SC No 99 two-speed, 7nhp, built 1910 and shown here at Andrew Bros, Terrington St Clements, Norfolk, in 1946. Fowells built 100 tractions at their small works from 1878 to 1922, seven having a third extra-slow speed of the Savage design.

Fowler SC No 3190, 6nhp, built 1877, two-speed with the replaceable fast-speed bell pinion. Shown thrashing at Cherry Tree Farm, Bellington, Buckinghamshire c1895. This is a delightful scene to the onlooker if not to all the workers, despite the pony-drawn beer barrel.

Garrett (Leiston, Suffolk) SC No 32890, 6nhp, two-speed, four-shaft, built 1916. Here at Youngsbury, Hertfordshire, in May 1950, is a stackyard end-of-thrashing scene in the last days of steam after 100 years' service—even the free-range hen has gone.

(*Above*) Marshall (Gainsborough) three-shaft design of 1880 but here fitted with a McLaren single cylinder and shown at Castor, Northamptonshire, c1900 when owned by A. Gibbons & Sons. Holding an oil-can was a common pose before the plate cameras (and the long exposures necessary). (*Below*) McLaren (Leeds) SC No 112, three-shaft, two-speed, probably 7nhp, built 1881, thrashing near Burmarsh, Kent, when owned by E. Checksfield of Burmarsh, in December 1944. The heavy Watt-type governor is in action and a pair of Salter safety-valve levers project from the regulator chest cover.

Ransomes, Sims & Jefferies (Ipswich) SC No 8980, two-speed, three-shaft and probably 5nhp, built 1889, here owned by W. Warren, Wrestlingworth, Bedfordshire. The pair of Salter safety valves were surmounted by an elegant brass vent. A later-date Pickering governor was fitted and the flywheel cast-in crank balance weight shows here at the top.

Robey (Lincoln) DCC No 29330, with inclined top slide valves and bored crosshead guides, 6nhp, three-speed, built 1910, shown here c1940 after thrashing. The rear wheel brake and the hose connected to the water-lifter will be noticed.

(*Above*) Armitage & Ruston (Chatteris, Cambridgeshire) was a very small concern which probably built only two tractions, this one c1880. The location was either Chatteris or Fowell's yard at St Ives. The final drive of the SC four-shaft design was by annular spur-rings on the rear wheels. The covered engine appears to be a Savage chain-drive c1873.
(*Below*) Ruston & Hornsby (Lincoln) No 163688 Class 'SH', two-speed, 7nhp, built c1931, thrashing in 1946 and owned by S. Adams, Timberland, Lincolnshire. The unseen reversing lever is forward for backward turning of the flywheel as in forward gear in a four-shaft engine.

Savage (King's Lynn) SC 'Sandringham' type No 388, 7nhp, built 1886 with Watt-type governor, twin Salter safety valves and Savage's patent extra-slow third speed. Like Fowells and earlier Fosters, the front axle was mounted well back. Shown at Buxton Lammas, Norfolk, September 1945.

Wallis & Steevens (Basingstoke) DCC 7nhp two-speed No 7496, built 1915, hauling a thrashing set of drum and elevator (an iron front wheel shows on the right) at Chartridge, Hertfordshire, in October 1945. There were two driving pins to the rear wheel and the governor was the Pickering type. Curiously the set was on the offside of the road and the engine carries the rampant horse of Aveling & Porter.

STEAM TRACTORS

The smaller versions of traction engines date from 1903 when an Act permitted tractions not exceeding five tons unladen (without coal and water) to travel on the roads at 5mph maximum in charge of one man. Makers quickly produced engines which met this weight stricture although medium-sized traction engines weighed about ten tons.

Early 'five-ton' tractors were fitted with single cylinders and two or three speeds. Virtually all the later tractors had compound cylinders with their greater steam economy and quieter exhaust.

Road speeds in the fast gear could at least double the legal maximum and hauled loads of ten tons were easily managed. Under average conditions these tractors could travel twelve miles with a load without taking up water and twenty miles without replenishing the coal bunker.

The apparent differences between traction engines and steam tractors (technically termed steam motor tractors) suited the latter for road work. The disc flywheel and cover plates to the motion enabled horse traffic to encounter tractors with slightly more confidence than when meeting general-purpose tractions. The extra boiler water tank, sprung axles, and often a canopy were further road work fittings. The obligatory fitting of solid rubber tyres to steam road engines from 1926 improved their performance, although the traditional cross-straked steel rear wheels were essential for soft going when gripping cleats could be fitted.

Duties of tractors were varied. Early designs included tractors weighing only three tons and an example is illustrated. Wallis & Steevens of Basingstoke were well-known for these small engines which could manage urban haulage including the vital job of pulling brewers' trailers. The small tractors had little reserve of steam, the driver was cramped and was obliged to stand at his work.

Timber-felling merchants favoured the 5-ton tractor because of its handiness, good power/weight ratio and the haulage facility of the winding cable. Indeed, large tree trunks could be loaded and transported in unfavourable winter weather on woodland sites.

A 1923 Act permitted 7½-ton tractors which proved to be the final designs of the purely traction engine arrangement. Rated at 4 or even 5nhp and rather higher geared than traction engines, these larger tractors, all with compound cylinders, were versatile engines. Boiler pressures were commonly 180 and even 200psi in the Burrells.

Showmen used several makes of larger tractors. They could haul a complete small roundabout set besides generating electricity for the galloper's lights.

On the farm 7½-ton tractors could belt-drive medium-sized thrashing drums having a grain output of about 2000lb per hr. The tractors' flywheels, being smaller than traction engines', required governed engine speeds of about 200rpm for thrashing.

Direct traction ploughing could be done by all except 3-ton tractors but soil conditions needed to be reasonably good. In any case, the newly imported 23cwt Ford internal-combustion tractor in 1918 superseded steam for field implement traction. An ingenious 'push and pull' steam tractor and plough is illustrated.

The classification 'steam tractor' includes those which combined design features of traction engines and steam wagons. Nearly all these were introduced from the mid-1920s to compete with petrol-engined haulage. The final drive to the rear wheels was usually by roller chain and there was considerable comfort for the driver. The designs were technically advanced in their day and Foden, Sentinel, Atkinson, Mann and Robey were the principal makers. Under- or overmounted engines which steamed from short locomotive-type or vertical multi-tubed boilers were used. In the designs using undermounted engines, these were virtually totally enclosed high-speed types using cam or poppet valve gears.

The Foden 'D' type tractor shown was the ultimate in a design which retained the traditional traction engine element. The short locomotive boiler worked at the high pressure of 230psi, steaming Foden's standard compound cylinders. The maker's special three-way valve passed boiler pressure steam direct to both cylinders by the driver-operated control. The steam was then correctly proportioned between the high- and low-pressure cylinders ensuring even and extra power for short periods of emergency use.

Roller bearings were fitted to the main components of the motion and other refinements included electric lighting, two sets of rear wheel brakes, Ackerman steering and a special cable-winding drum. A small modification to the cab-side enabled fly-wheel belt work to be undertaken. Unladen, the Foden 'D' tractor weighed 7 ton 19 cwt. The combination of undermounted engines and vertical boilers in steam tractors allowed good forward driving positions. The Sentinel was an excellent example of the undermounted type in which the driver sat at an angle to the forward boiler which was fired through a top chute. The engine was of duplex (or twin high-pressure) design, in which the cylinders shared the boiler steam. Camshaft-operated valves were used and the nominal maximum engine speed was 1500rpm. Again, the final drive to the rear wheels was by roller chain and all wheels were on pneumatics.

Garretts of Leiston, Suffolk, produced a unique design of agricultural steam tractor in 1917. It was intended for direct traction of field implements but for the reasons already mentioned the tractor came too late for general acceptance and only eight were made. The overmounted engine was of compound design using piston valves and worked at 200psi. This 'back-to-front' looking tractor could compete with petrol tractors on a cost per acre basis and indeed could undertake heavier work.

Generally speaking, steam tractors for road work hauled 10-ton trailers with considerable ease and could well exceed the 12mph legal speed limit. The Ackerman steering and good braking systems facilitated 'fast' road work, and in emergencies stops could be made by using the power of the engine as already explained in the traction engine section.

Aveling & Porter 5-ton DCC built c1905 and owned by Bomford & Evershed Ltd, Evesham, Worcestershire. It is fitted with special winch drum and tail-rope guide (at rear), probably for pond-dredging. The front wheels have extension rims.

(*Above*) Aveling & Porter 3-ton SC, shown when new at the Rochester 'Invicta' works (Rochester bridge is in the background) and almost certainly exhibited at the Maidstone, Kent, Show of 1904. Two speeds and a rear wheel brake were included. This three-shaft tractor could manage twelve miles between water stops. (*Below*) Aveling & Porter 6-ton DCC No 6898 with slide valves, built 1909 and owned by the Metropolitan Water Board, Hampton, Middlesex. The wheels are the final cast design. The fine acetylene headlamp and the upholstered sprung seat were non-standard fittings.

Burrell 5-ton DCC No 3851 with slide valves, built 1920 and here owned by E. Longhurst, Epsom and Dorking, Surrey. Shown standing and slightly sunk in near Ewell, Surrey, in April 1933. This is a three-speed fully-sprung tractor. Nowadays every brass plate and fitting would go from a standing engine.

Foden 'D' type DCC No 14080, with chain final drive and sold to HM King George VI at Sandringham on 30 November 1936. Modifications for sawbench work (governor gear, flywheel, cab and screens) and also overhauling, cleaning and painting were completed in sixteen days at the Foden Elworth Works, Sandbach.

(*Above*) Clayton & Shuttleworth SC slide valve No 36921, shown at the Stamp End Works, Lincoln, probably after testing when new c1904. This is almost certainly their first production 4-ton tractor and seen here in works grey before a sheet oscillated by workmen to avoid wrinkles showing in the photograph. (*Below*) Foster DCC with top slide valves, 'Wellington' type (Wellington Works, Lincoln) No 14378, built 1926. Standing at Hitchin, Hertfordshire, in 1946 when owned by Folbeg of Wyboston, Bedfordshire. The World War I tank motif signifies Foster's military work.

Garrett 'Agrimotor' or 'Suffolk Punch' field implements tractor introduced in 1918 showing its unique design of gear and chain-drive. This is No 33180, 1919, with 40bhp DCC piston-valve engine, locomotive boiler, superheater and economiser. Shown in 1944 at Easton when owned by J. Goddard, Tunstall, Suffolk.

(*Above*) Garrett 5-ton superheated DCC No 30988 with outside piston valves, built 1912. 516 Garrett tractors were made from 1907 to 1929, 44 of the mainly earlier ones being superheated. Here the boiler hand-hole plate is lying on the tank top when owned by G. & J. Cutting of Pettaugh, Suffolk, in 1948. (*Below*) Wallis & Steevens 4¼-ton DCC No 2927 with slide valves, built 1907, shown direct ploughing with 'push and pull' ploughs (the tractor is travelling in reverse and the ploughman is steering the engine). This is an 'oil-bath' design having an oil-tight motion casing. A small steam boiler feed pump is below the steering wheel.

Sentinel 'Standard' No 567 built in Glasgow by Alley & McLellan (first name of Sentinels) c1914 as a 6-ton undertype wagon. Rebuilt as a tractor with a Sentinel Shrewsbury-made front, it is shown in Liverpool in 1946 when owned by J. Blythe & Sons, Corn Millers, Liverpool.

PLOUGHING ENGINES

A few of the later and final designs of ploughing engines are illustrated. These were developed in a variety of forms from 1856 for ploughing by cable-hauling field implements using two engines as opposed to direct traction.

John Fowler of Leeds is synonymous with steam ploughing and his company achieved world-wide reputation from the early 1860s to the end of production in 1935. Other notable builders of PEs for the double-engine system were Aveling & Porter, McLaren and Burrell. All exported these engines and McLaren lasted longest, supplying Fowler spares after 1935.

During the last quarter of the nineteenth century and up to 1918, two-engine cable ploughing, successfully introduced by John Fowler in 1861, was widespread in the world. The World War I years caused the last upsurge in demand for steam-ploughing and cultivating in the United Kingdom. This work was mainly in contractors' hands because of costs; these were approximately £5000 for two compound PEs and complete tackle at that time. Ward & Dale Ltd of Sleaford, Lincolnshire, owned forty-eight PEs, twenty-four ploughs, twenty-four cultivators, water carts and living vans in 1908 and were probably the world's largest PE contractor.

Here the pictures show only one facet of steam cultivation. This subject is vast and may be studied from several specialist books.

Steam-cultivation demonstrations in recent years have shown the double-engine system to the rally-going public, and so the salient external features of the whole system—if not all the technicalities—may be seen at first hand. The massive construction of PEs and equipment has enabled examples of designs dating from c1870 to do their job today albeit with some re-building during the century.

In simplified terms, a pair of PEs are identical in design except for one cardinal point. One engine hauls and pays out its cable on the left or nearside while the other does so on the right side. Consequently the heavily-made cable guide arm projects from under each drum to the left or right. Opinions differ, but the PE with the left-pointing cable arm is known to the author as the left-hand engine.

The cable guide arm itself ending in two small pulleys—known as the 'monkey's head' because of the shape—rises and falls by the precise distance between the upper and lower drum coils; moreover this movement is gear-and-cam-timed so that each drum coil is guided next to the preceding one. This final design of guide dates from c1875.

The final horizontal-drum PEs shown derive the drum rotation via a bevel gear on the crankshaft meshing with a larger one on the short vertical shaft, a dog clutch on that shaft and finally low-ratio spur gearing to the gear ring of the drum itself. The clutch is operated by the driver's lever and of course engaged for hauling in.

A friction band on the drum is automatically applied when the slack cable is paid out. This slight braking action prevents a freely revolving drum from over-running its cable and displacing the unwinding coils. By using two sets of gears and a second clutch, two drum speeds were obtained on some of the later Fowler and McLaren PEs.

The standard high-carbon steel stranded cable is specially laid to withstand the severe abrasion by dragging over the land; ¾ or ⅞in diameter cables were used for the heavier duties. Standard lengths were 450yd, but 600yd could be fitted on deeper drums.

The whistle mounted on top of the regulator chest was an essential fitting for signalling to the distant engine that a pull could start. Given vision between the engines that was unobscured by a land rise or mist, then the driver ready to pull started when he saw the steam from the distant whistle.

PEs have two 'road' speeds, the slower being used in the field. The earlier engines needed a manual change of gear wheels, the one not in use resting on a vertical stub shaft on the footboard.

Generally speaking, PEs can carry out traction-engine work but were unwieldy and unsuitable for thrashing unless specially fitted with a governor.

The sporadic bouts of work up to full output required boilers of large capacity so that the fire did not require making-up during the actual pull. Firing, as well as feeding the boiler (by one of two injectors), was done when the PE was 'idle' between pulls.

The majority of the later Fowler compound PEs were 16nhp. Engines with slightly larger boilers were nominally 18nhp with a somewhat slower drum speed for ploughing at about 4mph. The faster drum speed was designed for cultivating at speeds up to 6mph but with a corresponding lesser pull on the cable.

A footplate view of a Fowler PE's motion work shows the most compact and powerful arrangement of the traction-engine type design. The crankshaft has three bearings and the two sets of link reversing gears for the compound cylinders work with the minimum clearance between them. The valve spindles project from the valve chest above and between the cylinders at a down-slanting angle. Hidden in the valve chest are the steam port faces to each cylinder. These faces not only slant downwards in line with the valve spindles but also incline inwards in a shallow V-shape. All this makes a very compact cylinder block which is entirely steam jacketed for greater efficiency.

Differential gear was not fitted to PEs although one late-1925 engine was so fitted. A rear-wheel centre pin had to be withdrawn when cornering at any sharp road bend, otherwise without differential gear the road surface could be torn by the rear wheels being driven at the same speed.

From 1881 onwards single-cylinder and the earlier double-cylinder PEs began to be super-seded by compounds. 'Singles' continued to be made and were favoured by some users, probably because of the less complex motion work. At work a 'single' PE emits staccato barks from the exhaust—a steam power sound which resounded over the fields.

Besides the usual five and six furrow double-ended plough, Fowlers made 698 other varieties. Overseas types included lava rock removers, deep vineyard ploughs and sugar cane trash buriers.

In 1859 Fowlers introduced a mole drainer whch in essence was a steel 'bullet' on the end of a deep fin. This wheeled implement was drawn by a horse windlass. The final mole drainers which exist today were drawn by the PE cables in a double purchase arrangement.

Double-ended ploughs for steam haulage by cable were introduced in 1856 in a joint design by John Fowler and William Worby of Ransomes & Sims, Ipswich, who built the first one. They were 'balance' ploughs because the multiple shares were balanced end-for-end about one central fixed axle and a pair of wheels. In use, particularly on land which was not level, the ploughing end tended to vary its depth in the soil even when two ploughmen were riding.

The 1885 Fowler invention of the 'anti-balance'

plough solved the problem. The central axle was given a limited endways movement which took place only when either of the two ends were cable-hauled. The effect was to weight the plough at the working end. At the finish of each pull the plough axle automatically travelled to its central and 'balance' position, thus enabling a manual pull-down of the plough end ready for the next haul. At best, this manhandling was very heavy work, a fact not always appreciated at today's demonstrations where there are a number of helpers.

A complete set of Fowler 16nhp PEs, plough, water cart and living van was farmer-owned up to 1932 very near the author's present home. The farm was mainly reclaimed marshland and one piece of dead flat arable land was reputed to be more than 500 acres without boundaries. This was ideal for steam ploughing.

This set travelling on the road was spectacular in sight and sound. The ninety feet of spaced-out machinery, completely steel-tyred and unsprung except for the living van, set up a clangour which could be heard a mile away. In all, the weight was about fifty tons and the twelve ploughshares glittered after a long bout of work.

A steam-plough team usually consisted of a foreman, two engine drivers, a ploughman and a youth for cooking. A standard living van had eating and sleeping accommodation for six men and was of course fitted with a coal-burning stove. Contractors' steam-ploughing sets often worked miles from their home yards and personal transport was non-existent for ploughmen.

Despite so much excellent work done by the double-engine steam-ploughing systems which either harrowed, cultivated or ploughed nearly 65,000 acres in the East Midlands alone during 1914, there were some complaints during that era. It was said that 'one engine is always idle'.

In an age of hard manual work, steam ploughmen often achieved the virtual impossible in sustained long hours and effort.

Aveling & Porter DCC No 8890 *Marshal Haig*, built 1918. The five-furrow anti-balance plough is nearly aligned for pulling by the unseen distant engine (A&P *General Byng*). Note that the engine cylinder block has outside valve chests.

Fowler-Goode tandem compound (originally a Fowler 16nhp SC) No 3938, built c1880 and one of five SC Fowler PEs converted during 1903 to 1918 by adding a low-pressure cylinder in front of the single cylinder, thus increasing efficiency. All work except the heavy machining was done by Edwin J. Goode and a few men at Lodge Farm, Elmdon, Essex. Shown when owned by Flack in 1946 at Chrishall, Essex.

(*Above*) Fowler SC No 2693, 14nhp, built 1886 with some rebuilding including a new chimney by the Oxford Steam Ploughing Co. 'Tank' steering is shown; the vertical steering wheel is behind the driver. The cultivator has reached the end of the pull in this 1973 demonstration in the ownership of Beeby Bros, Rempstone, Nottinghamshire. (*Below*) Fowler SC No 2050, 12nhp, built 1873, cultivating c1905 at Chenies, Buckinghamshire, when owned by T. T. Boughton. The Salter safety valve is blowing off hard through the ornamental casing. The slack cable (foreground) is shackled to the 'turn around' arm and goes to the distant engine.

Fowler DCC Class 'B4' No 8231, side drum, built during the short production period from 1899–1900. This design did not satisfactorily guide the cable at an angle under the manstand and make even drum coils. Seen here in 1949 and owned by J. Hosier, East Grafton, Berkshire.

Fowler DCC Class 'K14' No 13452, built 1914, named *King George V*. The drum speed was slightly higher than that of the 'BB' Class which immediately followed. Here a governor (behind the mechanical lubricator) is unusually fitted for belt work. The pulling cable is bar-taut and the drum coils are perfectly even. Shown at Coleshill, Buckinghamshire, in 1945.

(*Above*) Fowler DCC Class 'AA7' No 15257, 18nhp, built 1918, named *Repulse*, pulling in 1949 when owned by Dennis Bros, Kirton, Lincolnshire. The 'AA' Class was normally the largest PE used in the United Kingdom and designed for ploughing. Here the water tank is being refilled. (*Below*) Another working view of No 15257, deep ploughing with a Fowler four-furrow balance plough (three men have probably been on the ploughing end). The stubble is very long and the Lincolnshire location a bleak one in Autumn 1951.

Fowler SCs No 1071 (left) and 1072 (right); the pair were 10nhp and built in 1865. Known as the 'horizontal shaft' design because the cable drum was driven by two sets of bevel gears shown besides the drum pinion, they are owned here in 1936 by R. J. & H. Wilder, Wallingford, Berkshire.

Fowler DCC Class 'BB1' No 15154, 16nhp, built 1918 and still commercially used in the late 1960s. 'BBs' were the most numerous class in the United Kingdom and designed mainly for cultivating with a higher drum speed than for ploughing. Seen here in 1946 with a Fowler water cart and owned by H. C. Burgess, Pointon, Lincolnshire.

(*Above*) McLaren DCC No 1552, 14nhp, built in 1919 at Rye, Sussex, in 1936, owned by Bomford & Evershed, Evesham. The slide valves are outside the cylinders and the 'fast' crankshaft pinion is on the footboard stub shaft ready for manual changing. (*Below*) Fowler DCC Class 'BB1' No 15409 after a pull with a four-furrow anti-balance plough at Barks Farm, Walkden, Hertfordshire, in November 1955. After the distant engine starts to pull, the fire and boiler water in this one will be made up. The going is soft and spuds are on the rear wheels.

SHOWMANS ROAD LOCOMOTIVES

Fairground amusement caterers used steam engines for hauling equipment from 1880 to 1957 when a showmans engine was last commercially used in Birmingham. Probably the first recorded isolated use was in 1859 when three Bray traction engines were loaned to a showman.

The pictures show fairground locomotives in their final designs incorporating double-crank compound cylinders, front-mounted dynamos, full-length canopies and all the brass embellishments and decorative finish traditional to the fairground.

Essentially these engines, last made by Fowlers and Fosters in 1934, were heavy-haulage road locomotives with extra fittings suitable for handling and running the large round rides at the zenith of their development.

Invariably three speeds were fitted, sprung axles, pressed-on solid rubber tyres, a rear wheel rim brake and often a flywheel brake. The 'fast' gear and driving wheels reaching seven feet in diameter allowed surprising speeds even with heavy loads. On good level roads, 15mph averages were achieved hauling forty-five tons. This weight could be taken from Norwich to London in one working day before 1914.

It will be noted that showmans road locomotives (SRLs), like tractions, retained to the last the rudimentary steering arrangement of a straight swivelling front axle and low-ratio worm gearing plus chains. These mechanics were hardly suitable for main road traffic conditions during the later years of showmans engines. Surprisingly there appear to have been few accidents caused only by mis-steering.

The drivers had the poorest view of the way ahead among all traction types. The long low canopy, massive dynamo and the size and length of the whole engine were disadvantages that were not overcome by the second man—the actual steersman.

After 1910 when the first large electric scenic railway was introduced, a number of existing SRLs were specially fitted with a second, smaller dynamo (the exciter) and a tender-mounted long crane jib utilizing the engine's standard rear axle winding cable. The first showmans engine to be built with these fittings was Burrell's *Victory* in 1920 (see *Traction Engines and Steam Vehicles in Pictures*).

The eight passenger vehicles of scenic railways required such a heavy starting current that the front dynamo was insufficient. In practice, the belt was often thrown off. The exciter, belt-driven from the main dynamo, overcame this problem by using a complex electrical circuit in which the rail vehicles were controlled from the pay-box. In fact the eight peacocks, whales, dolphins etc never actually stopped but inched slowly forward at the end of each 3d ride.

The total maximum electrical output of a Burrell 8nhp 'scenic' engine was 29,700 watts (main dynamo) plus 8,800 watts (exciter), both at 110 volts. This exciter output was a momentary starting one and rapidly declined as the rail vehicles gained speed.

All this required nearly 45hp to be transmitted by the main 9in-wide driving belt and 'scenic' engines were fitted with a specially wide 9in flywheel rim compared with the normal 7in rim. The rear-mounted crane was used during building up and taking down scenic railways and the vehicles, weighing about 1½ tons, were lifted on and off the rails by this crane. It was a nice exercise of skill combined with the docility of controlled steam power.

Two 'scenic' engines were needed to run these large switchback railways. The second engine generated power for the multitude of lights, the large-capacity water pump for the 'cascade' opposite the organ and the blower for the organ.

This ensemble of steam hard at work was as fascinating as the glittering and blaring ride itself. Naturally pairs of these engines doing their best night after night attracted their own spectators. Apart from some real work at the Great Steam Fair at White Waltham in 1964, sustained engine outputs have not been seen for almost forty-five years. On the fairground only a few dim lights were fitted to engine canopies. The governed speed of 166rpm imparted a rhythmic to-and-fro rocking to the whole engine; the driving belt joint regularly slapped round the dynamo pulley while both dynamos emitted the peculiar singing hiss from their open brush gears and commutators. Above all was the steady sonorous thump in the chimney base from the exhaust beat of the compound cylinders taking steam at the maximum of 180 or 200psi. The drivers had little spare time and were either firing up, oiling round or attending to the water supply and boiler water level. The immediate surroundings were soon well wet and, despite the tall extension chimneys, the ground level atmosphere was pleasantly charged with coal smoke and the smell of steamy oil.

Showmans engines were made by the majority of traction builders but significant numbers came from fewer builders. Among the estimated 1000 plus engines used by showmen during the reign of steam, there were approximately 340 Burrells, 330 Fowlers, 154 Fosters and 52 Garretts, while McLaren, Wallis & Steevens, Foden and Aveling & Porter came next, all with less than 40. Eight further builders produced fewer than 10 each.

In common with tractions, SRLs were designed with either one or two intermediate gear shafts between the crankshaft and the rear axle. Burrells—having settled on the three-shaft design for the proven reason of less friction, hence more power at the rear axle—had the advantage of boiler space between the chimney and cylinders. This well suited the fitting of exciters.

Fowlers were four-shaft engines, therefore their cylinders were necessarily further forward from the rear axle. Exciter location was made somewhat difficult and a few engines had the second dynamo fitted to the boiler water tank top. One engine had a specially extended smokebox to allow room for the exciter between cylinder and chimney.

SLRs were made in 5, 6, 7 and 8nhp sizes but a few were rated 10nhp. Fowlers turned out their last four showmans in this size from 1932 to 1934. Burrell engines actually built as showmans included three rated at 10nhp although two of these had cylinder sizes identical with some of their last 8nhp designs.

Compound cylinders for dynamo-driving soon superseded singles and the few Burrell single-crank compound showmans. Obviously drivers could not easily manhandle a flywheel stopped on a 'dead centre' when it was belted to the dynamo and the facility of self-starting with compound cylinders overcame this disadvantage.

SLRs in the smaller nhp sizes mentioned were used for a variety of generating and hauling duties except the large 'scenics'. All the lighting on the later 'gallopers' and similar rides was supplied externally by a showmans engine. Only the rotation of the ride and its central organ was powered by the fixed steam centre engine. Round ride lighting was achieved by a heavy copper slip-ring attached to the top towards the centre.

Many showmans engines bore heavy cast-brass nameplates chosen by owners for appropriate, evocative and sometimes rather whimsical reasons. British reigning monarchs and members of the Royal Family were popular besides leaders of British and other Armed Forces. There is now a touch of history in *General Buller, Lord Kitchener* and *Marshal Foch*. G. T. Tuby of Doncaster uniquely recorded his rise in public office in that borough by naming his successive Burrell engines *The Councillor, The Alderman, The Mayor* and *Ex-Mayor. The Russell Baby*, a showmans tractor of 1920, was not named for its size but because of well-publicised litigation about paternity.

The heavy haulage capabilities of large SRLs have been mentioned in passing. Until 1896, nine towed vehicles were permissible and after that date three vehicles plus a water cart. All the earlier showmans wagons and living vans were carried on ordinary cart-type wheels of wood construction except for the iron tyres and brass-bushed hubs. Later, stronger 'artillery' wheels were used with metal hubs. The plain bearings of all these wheels often overheated and indeed this could set fire to the wheel. Transport of heavy fairground equipment was therefore beset by difficulties especially on the macadamised secondary roads before the early 1920s.

In hilly areas and where roads were bad (and this included steel engine wheels on granite setts or greasy wood blocks), the towed train had to be divided and the parts of each heavy vehicle such as the 5/7-ton organ truck cable-hauled by the stationary engine. Sharp corners sometimes caused a 'skidding round' operation on the heavier vehicles. Iron skid pans fastened by chains were placed under one or more leading wheels and the vehicle hauled bodily sideways.

It will be obvious that gear-changing on the engine was always effected before a hill for the reason given in the 'Traction Engines' section on page 7. An out-of-gear SLR on a slope with a 30-or 40-ton load was a preliminary to a disaster even if billets of wood were handy for the engine's rear wheels.

In the face of vicissitudes, the showmans road locomotives performed very well on and off the fairground. The engine drivers were capable, dedicated and sometimes dashing, handling their cumbersome engines weighing up to eighteen tons with sureness and ease. Another engine name was perhaps the most fitting—*King of the Road*.

Burrell DCC No 2804, 8nhp *The Griffin*, built 1906 as *The White Rose of York* for A. Payne of York. Seen in fine condition with the riding master and driver c1950 at Birmingham when owned by Pat Collins. This engine was one of the last four used on the fairground.

(*Above*) Burrell DCC No 4030, 8nhp special scenic, *The Dolphin*, built 1925 and the last Burrell showmans locomotive built at Thetford, shown generating on the fairground c1932 when owned by H. Shaw. Two mechanical cylinder lubricators (near the low-pressure valve chest) are unusual. (*Below*) Burrell DCC No 4021, 8nhp special scenic, *Lord Curzon*, built 1925. Shown with the second special scenic engine when generating for Harry Hall's 'Gorgeous Glittering Whales', the ornate top of which is dimly seem. The date is the late 1920s.

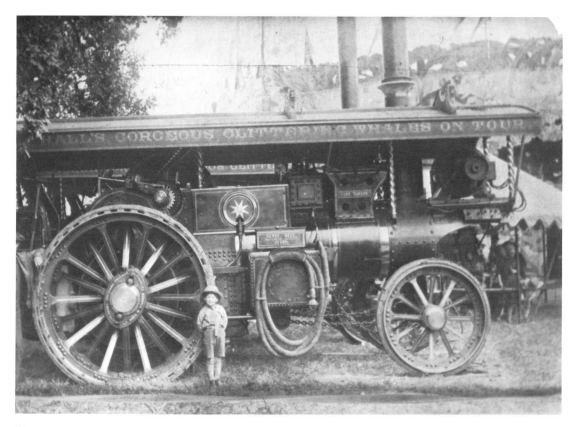

Fowler DCC Class 'B6' No 14424, built 1916 for the War Department as a heavy haulage engine. It was converted to a showmans engine for Pat Collins in 1921 as *Dreadnought* and then fitted with a rear crane. See here at Birmingham Onion Fair in 1950.

Fowler tractor 5-ton DCC No 14412, 3½nhp 'Tiger' type (as well as its name here), probably built 1917. Shown in a small fairground setting at Marks Tey, Essex, in June 1946 in the ownership of A. Downs, Peldon, Essex.

Fowler DCC No 15116, 8nhp *Bertha*, built 1917, generating for a 'whip' ride at Hounslow, Middlesex, in August 1947, when owned by Mrs John Beach & Sons. The driver's right hand is on the reversing lever, his left on the rear wheel brake handwheel. The rear vertical rollers were for guiding the winding cable.

(*Above*) Burrell tractor 5-ton DCC No 3497, 4nhp *Conquerer* (originally *May Queen*), built 1913, leaving Holm Green near Amersham, Buckinghamshire, in September 1950 when owned by Miss Sally Beach. The pneumatic-tyred front wheels were a modification. (*Below*) Burrell DCC No 2780, 8nhp *King Edward VII*, new to Charles Thurston of Norwich in November 1905 and used for his Royal Show (bioscope). Shown at Burrells, Thetford, in 1919 after fitting extended smokebox and exciter. The main dynamo and exciter are belted up.

Wallis & Steevens DCC No 7125, 7nhp, new at the Basingstoke Works in 1909. Built for Mrs J. Pettigrove, Charles St, Hatton Garden, London; an open-frame dynamo and rear wheel brake were fitted. The traditional excellent finish of all showmans locomotives is exemplified.

Wallis & Steevens DCC No 2643, 8nhp *Royal John*, new to Henry Jennings, Devizes, in 1904 (completed in 1903 for Boer War haulage but undelivered). Shown probably when new and profusely decorated; 'Electric Galloping Horses' refers to the lighting. The ride was steam-driven.

STEAM WAGONS

'Wagon' associates with horse-drawn vehicles but the term 'steam wagons' denotes road transport engines which carried their loads and in many cases could haul a trailer.

The concept of the steam wagon reasonably dates from Richard Trevithick's 1803 steam carriage. The first practically-used example was probably the special vehicle for hauling ships' boilers made in 1870 by John Yule, a Scotsman. There were at least thirty steam-wagon makers in the United Kingdom; these brief comments mention only a few of the principal firms.

Throughout their long production life ending in 1950 as exports, these steam vehicles were technically innovatory and in complete contrast to the staid traction-engine designs.

In broad terms steam wagons were classed as 'overtype' (engines mounted forward on top of locomotive boilers) or 'undertype' (engines below the load platform and front— usually vertical— boilers). There were many variations particularly in the earlier years from the 1890s. In 1907 Hindleys used vertical engines above the chassis; in 1924 Fowlers had V-type compound engines behind the driver and the Yorkshires had similarly-placed but in-line compound engines steamed from front transversely-mounted boilers.

Overtypes generally appeared from the early 1900s with the first Foden production design coming in 1901 after their experimental wagons from 1899. This company set the pattern for the vast majority of all 'pure' overtypes to the early 1930s. By the late 1920s, road steamers were dominated by undertypes.

The compact traction-engine boiler with its attached engine of the overtype was a well-proven, robust and reliable arrangement. The components were unenclosed and readily accessible. The starting valve in the compound cylinders could be used in Fodens at least so that only the high- or low-pressure cylinder worked. In cases of partial breakdown of the motion, this single cylinder working was a useful 'get-you-home' facility.

Locomotive boilers reached 230psi in the later overtypes (a considerably higher pressure than railway locomotives) and roller and ball-bearings were fitted. Invariably the final drive was by long roller chain to the rear wheels—a weighty arrangement at ten pounds per foot of chain. Earlier overtypes from Burrells had twin driving chains.

Two or three gear-wheel speeds combined with different slide-valve settings, obtained by positioning the reversing lever, provided overtypes with power variations for any conditions.

Early overtypes had steel-straked or plain steel tyred wheels and drivers' accommodation was spartan. Indeed Mann and Foden wagons provided a side perch for the steersman/driver whose view ahead was severely restricted and then to either the near or offside.

Early braking was by friction bands on the rear wheels together with the engine itself when the regulator was closed as for normal stopping. In emergencies, the reverse lever could be moved directly from forward to reverse position while the steam regulator remained open (see page 8).

The legal speed limit for overtypes on solid rubber tyres before the early 1920s was 12mph and 5mph with a trailer. These figures were usually considerably exceeded; a 1908 Foden overtype could *average* 12–13mph unladen with a trailer and 9mph with a carried and towed load totalling 10½ tons. These figures applied to a steam wagon of a nominal 5-ton capacity on journeys which included hills taken in low gear. Generally speaking, pre-1914 five-tonners travelled 20 miles per hundredweight of coal, up to 9 miles on one firing of the boiler and needed to stop every 22 miles maximum for taking water.

Four-wheel, 5/7-ton overtypes remained in general production until the mid-1920s and as late as 1933 in the case of Fosters. The basic overtype design remained unchanged to the end but all makers improved boilers, braking (steam and hand brakes) and particularly the cabs. These became completely protected and screened with 'upholstered', if hard, seating for drivers and mates. Pneumatic-tyred wheels appeared on Robeys, Fodens and a few Garretts.

Overtypes of larger capacity, either flexible (articulated) or rigid six-wheelers of nominally 10/12-ton payloads carried on the wagon itself came in the 1920s. These were Taskers and Ransomes (two only from the latter) in 1922, Robey in 1925 and Foden in 1928.

The final overtypes worked at 15–18mph on the level and as usual were often grossly overloaded. Many wagons were mechanical tippers, end-tippers being more numerous than three-way. Their progress was marked by the rapid muffled thudding of the exhaust from the compound cylinders and the peculiar metallic clicking of the rear chain over the sprockets.

Probably very little tangible history of road steam remains unrecorded during recent years. Nevertheless in 1977, while the author was examining and listing nearly 11,000 engineering drawings dating from 1870 from Alfred Dodman

Ltd of King's Lynn, general arrangements and details of their four-wheel overtype dated 1914 were found. Apparently owing much to the Clayton & Shuttleworth engine, the Dodman almost certainly was never made—probably due to World War I.

In turning from the overtype to the undertype, notable 'in between' layouts included the Yorkshires and Fowlers mentioned in passing.

Yorkshire was the third largest steam-wagon maker after Foden and Sentinel. Three main Yorkshire designs appeared from 1901 to 1937. Throughout they used a centrally-fired boiler having the drum transversely at the very front. Technically, the boiler was a locomotive firebox with water and return tubes within each side-projecting half of the drum. Their early wagons had undermounted compound engines with each cylinder attached below the chassis frame and all-gear drive. The final drive was later altered to roller chain. From 1913, Yorkshire engines were vertical compounds situated immediately behind the driver. Hackworth valve gear was fitted with fixed cut-off positions obtained as usual from movements of the reversing lever. Final Yorkshires had three gear speeds and their first shaft-drive came in 1922 followed by a shaft-drive rigid six-wheeler in 1928. Their last three wagons were shaft-drive four-wheelers made up to 1937.

The final Yorkshire six-wheelers were on pneumatics and had separate shaft drives to each rear axle. These fine wagons had published payloads of 10/15 tons, carried 300 gallons of water and had a coal capacity sufficient for approximately 120 miles using a hundredweight per 16/17 road miles.

A Fowler shaft-drive steam wagon was the sole and rather late design from this eminent builder of traction, showmans and ploughing engines. The wagon had a front vertical water-tube boiler, top fired by chute. The engine was a 90° V-compound, mounted behind the driver and above the chassis.

A 1924 test run of this wagon showed that for 86 miles the average speed was 19.1mph with a load of 6½ tons. One hundredweight of best Welsh steam coal was used per 17½ miles and the 100 gallons of water carried in the tank were sufficient for about 30 miles travelled.

Pure undertypes were dominated by Sentinels since their first design of 1906. The only Sentinel overtype design appeared in 1911–12 when sixteen were made. There were five development stages of undertypes, all with vertical front boilers and the first four designs used duplex totally-enclosed horizontal engines with cam-operated poppet valves.

The penultimate Sentinels were the 'DG' wagons of 1927–32. These were double geared in four-, six- and eight-wheeled types, all retaining roller chain final drive. The boilers worked at the high maximum of 275psi and average payloads of the DG6 and DG8 were 12 and 15 tons respectively. These weights could be exceeded to maxima of 15 and 20 tons. The imposing DG8 illustrated appeared in 1930; its length was 29½ft and the four steering road wheels were a first for Sentinel Ltd. All the later DG8s had pneumatics on their twelve wheels as optional to solid rubbers.

The Sentinel 'S' type (shaft drive) 1932–39 (for the UK) and c1946–50 (100 S6 types to the Argentine) is generally agreed to have been the most advanced production steam lorry although the final Fodens, when well-driven, could equal it. Made in four-, six- and eight-wheeled versions, the latter rigid with four steering, all the 'S' range were on pneumatics. The engine was the same throughout and a departure from Sentinel's traditional duplex.

The 'S' engine was single-acting having four in-line cylinders and the cam and poppet valve gear was retained. The engine with integral two-speed gearbox was set across the chassis so that the crankshaft was in line with the cardan shaft drive. The boiler pump, dynamo and air compressor for tyre inflating were attached to and driven by the engine which, with all these fittings, weighed 1008lb. The maximum brake horsepower was 124 at 800rpm. The vertical cross water-tube boiler working at 255psi and, fitted with a superheater, was immediately behind the driving position. This arrangement provided the optimum forward view in a comfortable completely protected cab. This was conveniently fitted with illuminated dashboard instrumentation.

Optional boiler extras were automatic firing by engine-driven Archimedean screw and automatic water-level maintenance. In practice, these extras meant that drivers only needed to make minor and infrequent adjustments of the boiler water feed. This boiler, extras and complete fittings weighed 1980lb. Raising full steam pressure from cold took 1¼ hours (always a *bête noire* with boilers containing a significant volume of water). The maximum payloads of the 'S' type variations appeared to be for non-tippers: S4—7.5 tons, S6—11.9 tons, S8—13 tons.

After 1945 only S6s were made and fitted with hydraulic brakes on all wheels whereas the earlier 'S' types had manual and steam brakes to the rear wheels only.

Fully-loaded 'S' types could travel about 57 miles between water stops and about 190 miles on the 5 hundredweight of fuel carried in the bunker. In all, their performance was well ahead of petrol and diesel lorries of the 1930s. Indeed, these

Sentinels could match internal combustion lorries for twenty years or so after the end of production of steam wagons in the United Kingdom.

Garretts produced the first of all rigid six-wheel undertypes in 1926, having already turned out 689 'pure' overtypes from 1909 to 1927. This undertype had a 12/15-ton capacity and was driven by a duplex piston-valved engine fitted with Joy valve gear. The final drive was double chain from the engine sprocket and from the leading rear axle to the trailing one. The wagon was vastly improved from 1927, the engine becoming a poppet-valve design with sixteen valves for the two cylinders. This design was repeated in all the other 114 Garrett rigid six-wheelers to 1931, two of the last four being on pneumatics. The chain drive remained to the last.

The 1926–7 Garrett undertypes as described suffered from some frame fracturing besides breakages of the front stub axles. After redesigning, a number of the later wagons lasted well in service and could average up to 23mph for travelling times of approximately 7 hours.

Foden's first undertype ('E' type) appeared in 1924 with the almost usual duplex cylinder poppet-valve engine steamed by a vertical water-tube boiler supplying superheated steam at a maximum of 250psi. The 'E' types were single speed with shaft drive. Forty-three were made including seventeen rigid six-wheelers. Unfortunately, and perhaps because of their weight, these wagons did not sell at all well and a few were even broken up unsold. The foresight of preservation value had no part in the working era.

Doubtless the 'E' type performance could be outstanding. A skilled driver has recorded some facts noted in 1928 when he was driving an 'E' type on the solid rubbers fitted to all the class. The wagon was loaded with 21½ tons of steel plate plus 1½ tons of boiler coal, the total of vehicle and load being 30¾ tons. At one point on the level during a long journey, the wagon's speedometer recorded 52mph. At this exciting moment the Law intervened. The same journey included a long hill having a part inclined at one in six. The hill work was easily managed on low gear.

Despite the fast-growing internal-combustion competition, road steam makers continued their supreme efforts. The inherent economical drawbacks of steam-raising time, vehicle weight and boiler maintenance were now added to by axle-loading taxation which *per se* was an unfair blow against the steam wagon. Its excellent road performance remained clutchless, non-polluting and gearless for starting purposes.

The Foden 'O' and 'Q' types were introduced in 1930 after one experimental and successful wagon was developed from the 'E' type. The 'O' type or Speed-Six and 'Q' type or Speed-Twelve (named for their payloads) ceased in 1932 and ended the steam era at Fodens. The last actual sale was in 1935 when a 1931 Speed-Six was sold to a German steam transport company.

It would seem probable that Foden's last steam wagons were insufficiently road-tested because of the harsh economic conditions. In the hands of skilled drivers, outstanding performances were obtained both in speed and silence. One eight-year-old Speed-Six, reboilered, worked for twenty years and could exceed 55mph loaded at the end of its life. Another Speed-Six usually worked day and night for six days a week.

Unfortunately, complaints about heavy water consumption, excessive blowing off at the safety valve and costly boiler maintenance were more numerous than comments of satisfaction.

Fodens fitted one experimental front condenser in an attempt to lessen the water consumption, but the boiler water became contaminated with oil and the idea was dropped.

The first type of boiler fitted to the 'O' type was a horizontal water-tube variety but at road speeds above 20mph it failed to supply enough steam. The second and final boiler was the same in outline as an 80° sharp bend in a domestic circular drainpipe. The firebox and shorter end was forward and vertical; the longer portion therefore sloped backwards at a 10° rising angle. This part contained ninety-four water tubes arranged in rows intersecting at 30°. The boiler working pressure was 275psi. Firing was by top chute at the driver's left.

'O' and 'Q' engines were identical and had duplex cylinders with cam and poppet valves. The engine and integral two-speed gearbox was under the chassis and very close to the boiler. The gearbox shaft drove direct to a rear-axle worm gear via an internal combustion type of cardan shaft; in the six-wheel wagon this shaft and the axle gearing were duplicated, thus driving both rear axles.

All revolving shafts and road wheels were fitted with roller and/or ball bearings. Internal-expanding steam brakes operated on all wheels and a manual brake was provided for parking.

In practice the engine was capable of a very high performance—indeed, beyond its design intentions. At 45mph the engine turned at 1590rpm and the vehicle could reach 60mph loaded. At the higher speeds valve failure could result but the general circumstances precluded a re-design to avoid this.

In the early 1930s the author often saw Foden's only example of these types supplied in Norfolk. No 13838 was a Speed-Twelve; it appeared to 'ghost' along and overtake any grinding internal-combustion lorry of the day.

An immaculate fleet of Allchin 5-ton overtypes owned by Locket & Judkins Ltd, Fenchurch Avenue, London. Four were supplied new in 1916–19. The wheels are wood and steel. These Allchins were first produced in 1916 with compound engines and chain drive to the rear axle.

Commercial steam in full cry: Atkinson undertypes rebuilt as bulk grain carriers from tractors. Works No 19 built 1917 (left) and Works No 25 (right) shown at Liverpool in 1946 and owned by Bibby & Sons Ltd. Note the waiting work horse on the right.

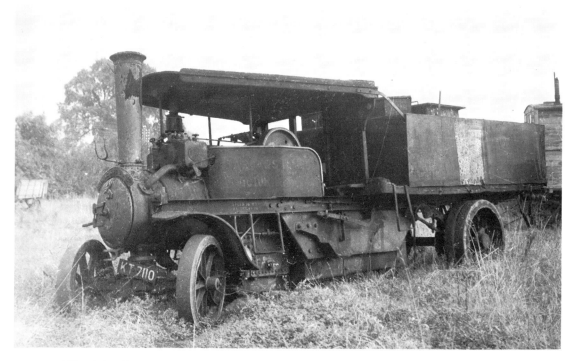

(*Above*) Aveling & Porter compound 5-ton overtype with Ackerman steering: No 8687, built c1915 and new to Budden & Biggs, Strood, seen here in 1935 when owned by G. Taylor, Redbourn, Hertfordshire. (*Below*) Clayton & Shuttleworth 3-ton overtype wagon, mechanical end-tipper No 48683, built c1919 on composite wood wheels with a locomotive boiler of 200psi working pressure, compound cylinders and chain final drive. Owned by Joshua Smith, coal merchant, Bradford, Yorkshire.

(*Above*) Hindley (Bourton, Dorset) 5-ton manual tipping wagon built c1907: this second design had a compound enclosed vertical engine and locomotive boiler with a high circular firebox. Shown new at the Dorset village works. (*Below*) Hindley 5-ton wagon in pastoral Bourton c1906. This is probably the prototype under test using the wagon's winding cable anchored left. The boiler design is clearly shown.

Burrell 5-ton overtype, the first production design with compound cylinders, three speeds, locomotive boiler working at 200psi and final drive by double chain. Built 1911 or 1912 and shown (possibly at St Albans) with a remover's container load and trailer.

Clayton & Shuttleworth water sprinkler and sewer-flushing wagon No T1092, a compound overtype on a standard 5-ton chassis and new in 1920. The tank held 1000 gallons and the front sprinklers were independently controlled from the cab.

(*Above*) Foden 5-ton compound overtype made 1910–25 with modifications. The legal speed limit was 5mph; the driver is in the normal but precarious position and an ordinary farm wagon forms the trailer. Seen here c1912; location unknown. (*Below*) A similar Foden wagon shown c1920 at Bourton, Dorset, when hauling a new boiler from S. Hindley & Sons' Works. Part of a standard four-wheel traction trailer is shown.

Foden 'E' type undertype No 13180 delivered in February 1929. This penultimate design was produced from 1924 to 1930 but first delivered in 1926: it had duplex cylinders, shaft-drive, differential rear axles and hydraulic brakes. The unladen weight was 10 tons and the nominal payload 9 tons, but 21 tons could be carried. (*Courtesy Fodens Ltd*)

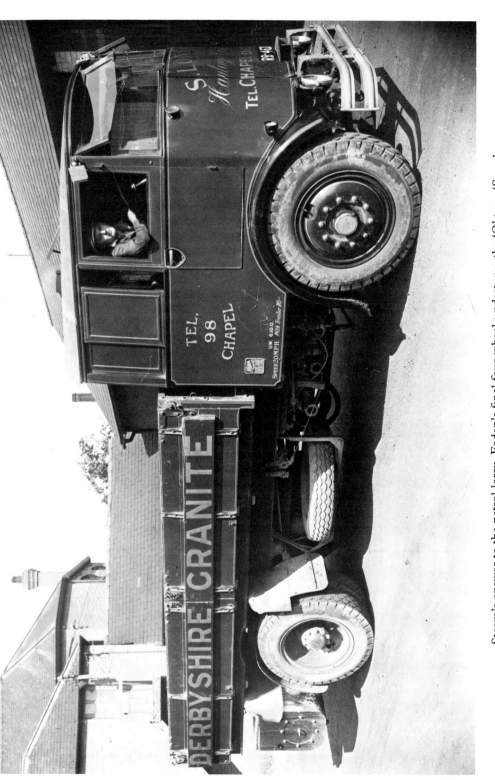

Steam's answer to the petrol lorry. Foden's final four-wheel undertype, the 'O' type 'Speed-Six' with totally enclosed transverse duplex engine, integral two-speed gearbox, shaft drive and locomotive cum water-tube boiler working at 275psi. Its official speed was 45mph; its possible speed 60mph. This is No 14028 built in 1931. (*Courtesy Fodens Ltd*)

(*Above*) Garrett's final undertype design, 8-ton three-way hydraulic tipper built 1931, No 35454, shown working at Wandsworth Gas Co, London, in 1947. The vertical boiler worked at 250psi and the duplex cylinder engine was fitted with cam-operated poppet valves. (*Below*) Leyland (Yorkshire) 8-ton wagon No F2/72/16822 built in 1926 and the final design. The engine had duplex cylinders with cam and poppet valves and integral two speeds; roller bearings were fitted to the crankshaft and transmission. It is seen here working in Liverpool in 1947 as Tate & Lyle Ltd No 2.

Londonderry (*The Marquis of Londonderry*, Seaham Harbour, Co Durham) undertype No 38 built c1906 and new to the North Eastern Railway. It was fitted with Ackerman steering, sprung axles and vertical boiler working at 200psi, and the totally-enclosed compound engine was fitted with slide valves. Shown at York in 1915.

(*Above*) The first of two designs of the 'Super-Sentinel' steam omnibus with a thirty-two-seater body built by E. & H. Hora Ltd, Peckham, London. A 'Super-Sentinel' wagon chassis was used with the standard duplex cylinder engine and spiral tube vertical boiler pressed to 230psi. Water stops were necessary about every thirty miles and a maximum (illegal) speed of nearly 40mph was possible. Seen here at the Leicester Show in 1924. (*Courtesy J. L. Thomas*) (*Below*) Sentinel type 'DG8' No 8568 built in 1931 and the last of the eight rigid eight-wheel types made, all with tipping bodies. The engine and boiler (275psi) were improved versions of the 'Super-Sentinel' designs. The 'DG8' was the first of its type produced and the four steering road wheels were Sentinel's invention. The maximum nominal payload was 12.05 tons. (*Courtesy J. L. Thomas*)

Sentinel 'S-4' shaft-drive No 9227 built 1934 and seen with a good load at Nine Elms Lane, London, in 1947. This final design had a four-in-line single-acting engine, an integral two-speed gearbox and four positions of the camshaft operating the valves. The listed maximum payload was 7½ tons and maximum speed about 45mph.

A Mann (Leeds) 5-ton overtype wagon reposing in the ownership of Surbiton District Council, Surrey, in 1935. The compound engine had single eccentric radial valve gear. The locomotive boiler had side-firing and worked at 200psi. These wagons were produced from 1913 to 1929 when Manns ceased steam wagon work.

(*Above*) The first Savage (King's Lynn) wagon made, class OA, owing much to C. & A. Musker of Liverpool. The boiler was a Musker water-tube design and the undermounted compound engine was fitted with piston valves. The final drive was by twin roller chains. Shown here probably when new in 1903. (*Below*) Wallis & Steevens (Basingstoke) 2-ton delivery van (second type). This delightful vehicle, built c1913, had a short locomotive boiler and probably a compound vertical engine with ordinary slide valve gear, the transmission being via spur gears and final chain. An LSWR van appears in the background.

Yorkshire (Leeds) type 'WG' double-chain two-speed six-wheel flexible wagon No 2181, new to Manbre & Garton Ltd, sugar refiners, of Hammersmith, London, in 1930. The transverse boiler steamed a vertical compound engine behind the cab. Joy valve gear was fitted and 100+ maximum bhp was produced at 1000rpm. The payload was 10/12 tons.

STEAM ROLLERS

The distinction of being first to use a steam engine for road-rolling is held by William Bray of Folkestone, Kent. In 1859 he used his traction engine to tow a separate heavy roller near Hyde Park, London. The only known photograph of this engine with its inventor/ builder in 1859 is shown on page 11. From 1860 to 1865 eight purpose-built steam rollers each weighing fifteen tons and of French design worked in Paris; some of these were built by Manning Wardle & Co of Leeds, Yorkshire.

In 1865, too, the famous name of Aveling (later Aveling & Porter) first came to notice. Thomas Aveling of Rochester, Kent, produced his first steam-rolling arrangement of his traction engine and a 15-ton towed roller. This was made for him at Erith, Kent, and was a section of cast-iron bridge pier 10ft in diameter and 9ft wide fitted in a massive wood frame. This cumbersome ensemble was superseded by Aveling's first complete steam roller in 1866.

Roads in the United Kingdom during the 1860s were in a very poor state caused by neglect. Their previous excellence during the horse-coach era had gone because of the fanatical zest for the new railways. In 1865 road speeds were legally restricted to 4mph in the countryside and 2mph in towns. Indeed, steam transport was banned in some towns and even villages or restricted to the dark hours. Sporadic road 'repairs' of the time consisted of merely spreading loose stone fillings for the vehicular horse traffic to consolidate. The suffering of horses on such freshly-stoned surfaces was hardly considered. In such conditions the need for efficient road-rolling was apparent to Thomas Aveling if to no-one else. His line of at least 8,000 steam rollers became well-known in the United Kingdom and many other countries. The very last, made by Armstrong-Vickers, Newcastle, as sub-contractors to Aveling-Barford, Grantham, Lincolnshire, were exported in 1950 to Siam and Indonesia. There, growing timber for fuel was always available.

Apart from one early and curious design, the majority of the pictures show steam rollers which can be recalled in their working years. Clearly these rollers are basically traction engines with a front roll instead of wheels. Several makers produced 'convertibles' in which either a front roll or standard traction wheels could be fitted by means of a bolted flanged ring in front of the smokebox.

The majority of traction builders included rollers in their production range. Generally the nominal weights of standard compound three-roll engines offered were from four to sixteen tons in two-ton steps. These weights were without coal and water and could vary by as much as two tons from the nominal weight figure.

Aveling, Aveling & Porter and Aveling-Barford were foremost as steam-roller builders, having produced nearly two-thirds of the approximate total of 6250 for United Kingdom use up to 1950 (nearly all had been made by 1930). The other approximate one-third was produced mainly by Fowler, Marshall and Wallis & Steevens. There were, therefore, comparatively few steam rollers made by some other well-known traction makers—Burrell, Allchin, Robey, Tasker, McLaren, and Clayton & Shuttleworth.

Obviously a prime necessity of a three-roll engine is that the rolling widths, front and rear, should overlap only sufficiently for a slightly curved track to be covered. The front roll design varied over the years. Aveling first used 'split' conical rolls in 1871 in an attempt to suit the universal camber of roads. By 1878 nearly parallel rolls were used but the rear ones were usually made some ¾in less in diameter on the insides to lessen or avoid wear caused by road camber. Aveling made specially large diameter (nearly equal to the rear) front rolls in 1912 for France but this feature which produced a more even surface compaction was not taken up in the United Kingdom until 1921.

Tandem rollers having a full-width roll at front and rear were an important variation from standard three-roll designs. A few Robey tandem rollers were rebuilt as 'tri-tandems' with dual rear rolls. Central vertical boilers were used in some tandem rollers in which duplex cylinder engines and special valve gears enabled instantaneous reversing.

The principal differences between the standard three-roller mechanics and traction engines may not be immediately obvious. Rolling speeds of 1½–2mph required low gears in each of the two speeds. The mainly straight-line work did not need differential gears and they were infrequently fitted except in the Marshall 'Universal' and Wallis & Steevens 'Advance'. Roller-steering was low-geared using a worm wheel of large diameter. Rapid cornering was unnecessary and there was considerable friction in turning the front roll although it was made in freely-rotating halves. Crankshaft eccentric-driven boiler feed pumps were invariably fitted, either direct drive or reduction geared.

In 1886, Aveling & Porter first used compound cylinders for steam rollers. The in-line or tandem arrangement of the cylinders was then used although the normal side-by-side compound cylinders soon followed.

Burrells first used their renowned single-crank compound cylinders for a roller in 1891. A total of 163 SCC Burrell rollers were made, the last in 1926, and thirteen were four-shaft design. The greater part of this production went to Saxony, Germany and France but a 17-ton and a 16-ton SCC were exported to Italy. The quieter exhaust of all compounds was some advantage in close proximity to horse-traffic.

Until the early 1900s, steam rollers had solid cast-iron rims or rolls and when these wore thin, complete wheel replacements were necessary. Renewable rolls or plates were offered from about 1910 although re-plating appears to have been done from c1905.

Small but important steam-roller accessories were spring-loaded or weighted roll scrapers and small bore piped water for roll wetting.

In pre-pneumatic drill times, steam rollers fitted with heavy rear-axle attachments could tear up any tar/stone road surface. Among several makers of these road scarifiers, Morrison, Allen and Price were widely used and could scarify in forward and reverse directions. The initial 'bite' in scarifying was made by setting the tines in a hole which was hand-pickaxed into or through the consolidated surface. The scarified widths were about equal to one rear roll width.

In 1920 a new series of Aveling & Porter three-roll machines were introduced. Their wide range indicates the variety generally offered in this type of steam roller alone. These A&P rollers ranged from six to twenty tons in nineteen varieties including single and compound cylinders with piston valves replacing the traditional slide valves. The former were both cheaper and easier to make besides effecting a claimed efficiency gain of approximately 8% because of lower working friction. Piston valves had one disadvantage which customers never overlooked. Downhill work or an overfull boiler sometimes caused trapped water in the cylinders. Slide valves can lift and release water but its only escape from a piston-valved cylinder is to burst the end cover/s. Special spring-loaded cylinder release valves were later fitted but customer resistance lingered which gravely affected Aveling & Porter's last years in steam.

The return of the slide valve came in 1930, only to be replaced yet again by piston valves in the final Aveling-Barford rollers built at Grantham, Lincolnshire, to designs by Ruston & Hornsby of Lincoln.

The tandem type of steam roller came in 1902 in an attempt to meet the difficult rolling requirements for hot bituminous surfaces. These needed long strip-rolling and only tandem rollers could deal with widths between tramlines.

Normal traction types of engine cannot reverse quickly which is essential on asphalt materials to prevent sinking. In 1911 Aveling & Porter's tandem roll with a side-mounted vertical engine directly bevel-geared to the rear roll improved the 'quick reverse'.

This problem was solved by tandem rollers with centrally-placed vertical boilers and horizontal duplex cylinder engines (aptly named 'coffee-pots'). A special single-eccentric valve gear named the Klug gear combined with the absence of a fly-wheel enabled these rollers to reverse instantly.

The 272 Wallis & Steevens 8/10-ton 'Advance' three-roll type introduced in 1923 were also designed for instant reversing. Their duplex piston valve cylinders, 90° cranks and no flywheel, together with gear quadrant steering and a differential, all made their name well-deserved. In addition, the rear axle was in halves and adjustable to suit various camber angles. Their weight distribution was evenly disposed between the three rolls. The 'Advance' remained in use until the end of commercial steam rolling.

The unique 3-ton Wallis & Steevens 'Simplicity' roller illustrated resulted from a 1925 patent. It was not a saleable success despite its boiler angle which ensured a water-covered firebox, and only fifteen were made. Earlier 3-ton rollers from Wallis & Steevens were of conventional design but, like their small tractors, proved somewhat difficult to drive for prolonged periods.

Marshalls of Gainsborough, Lincolnshire, made 'quick-reverse' rollers in tandem and three-roll versions. Their final design—the 'Universal'—was three-roll in which the rolls were water-ballasted for weight variations. Introduced in 1926, the 'Universal' had duplex cylinders and Marshall's radial valve gear giving a quick reverse. Precise adjustment of the rear rolls for camber, differential gear and direct gear steering were other refinements. By this time steam roller sales were greatly diminished and only eighteen 'Universals' were sold in the United Kingdom.

The last years of steam rolling after nearly a century of service are still remembered but, together with all these other steam prime movers at work, will soon be beyond recall.

Travelling steam rollers of county councils and contractors were ubiquitous on the highways and rural byways of the British Isles. The roller, living van and water-tank cart drawn up on a grassy patch after the day's work were a tranquil and benign sight. It was comfortably warm standing near the boiler and only a quiet sizzling denoted a readiness for the next morning. The ashpan damper was closed and often a metal sheet covered the chimney top. An appetizing smell came from the living van and a wisp of coal smoke ascended from the stovepipe.

Should the roller have been an Aveling & Porter, the brass rampant horse surmounted the 'Invicta' scroll. 'Unconquered' well-described all steam rollers and reflected Thomas Aveling's faith and foresight when he introduced his insignia in 1865.

Allchin SC 10-ton No 1131R, built in 1900 and Allchin's first roller. Seen here in perfect external condition when owned by the Borough of Northampton in April 1947.

Armstrong-Whitworth (Openshaw, Manchester) DCC with piston valves and fitted with Morrison scarifier. These rollers were produced from c1921 to 1930: as made, the piston valves design caused severe steam loss. Roll water sprays are fitted.

A picture unchanged for half a century. Aveling & Porter No 12545 built c1927, SC piston valve and fitted with a Morrison scarifier, takes the living van and water cart up a slope near Billington, Bedfordshire, in October 1954. The safety valves are lifting. The familiar sight has not attracted the baker's roundspeople.

Aveling & Porter SC 12-ton slide valve No 3798, built 1896 and fitted with Morrison scarifier. Shown as No 7 when owned by the Essex Steam Rolling Association. The water-cart hand-bucket pump shows behind the horse driver: photographed c1912.

Aveling & Porter convertible SC No 1262 built 1876, shown when owned by T. T. Boughton, Amersham, Buckinghamshire, c1898. The front roll had just been fitted in place of the traction wheels; later, the engine was changed back to a traction. A Watt-type governor is shown.

(*Above*) Aveling & Porter convertible SC No 2527 built in 1889 and shown c1898, also owned by T. T. Boughton. The traction axle swivel pin is seen below the smokebox. The very high roll-head casting appeared on the earlier A&P rollers. The safety valves are blowing off and a rural four-way signpost shows behind the driver. (*Below*) Brown & May (Devizes, Wiltshire) No 8130, a three-shaft, two-speed roller converted from a tractor built 1910, probably 8-ton as shown here at Billingshurst, Sussex, when owned by Carter Bros. The roll-head casting and the steering-chain shock springs are unusual.

Burrell No 4083, SC 8-ton, three-shaft with a Price two-tine scarifier. This was the last Burrell roller made, completed in January 1928. Shown working near Blythborough, Suffolk, in June 1948 when owned by East Suffolk County Council.

(*Above*) Clayton & Shuttleworth SC 10-ton No 34380, built 1902, a year after their first production roller. Shown, probably when new, with a Bomford scarifier and owned by Stranraer Town Council. The roll pivot is at the very front of its casting and a fine set of oil lamps is in place. (*Below*) Fowler SC No 15984, built 1924, seen after an accident with a tram at Nine Elms Lane, London, in June 1947. Owned by the Mechanical Spraying & Grouting Co, this roller has a tar-sprayer on the boiler tank, chain-driven from the crankshaft sprocket. LBC on the refuse bin is Lambeth Borough Council.

Garrett (Leiston, Suffolk) light DCC 10-ton No 34084, built 1922, working in June 1948 as No 3 for the East Suffolk County Council. The steering chains are on extended 'cotton-reels' and the roller is extremely well-kept: location unknown but the Gainsborough-like oak trees are typical of Suffolk.

(*Above*) Green (Leeds) SC No 1657, a design introduced in 1910 with offside valve chest. A total of 200 Green rollers was made for the United Kingdom between 1872 and 1931. Shown near Boston, Lincolnshire, in 1935 when owned by J. Dickerson & Sons (Elmley) Ltd, near Doncaster, Yorkshire. (*Below*) Millar-Marshall tandem, probably 10-ton and made in the early 1920s in seven weights from 4 to 10 tons with duplex cylinders and Marshall's radial valve gear giving instant reverse for asphalts. Steam-steering gear operated direct-quadrant steering—a total arrangement unsuitable for inexperienced drivers!

Marshall Class 'S' 12-ton No 79260, built 1925, working at Dunfermline, Scotland. Marshall's radial valve gear and piston valve were fitted. The rear roll spokes were a one-piece steel casting. The 'S' class were made from 1923 to 1944: from 1948 1000 were built in India.

McLaren 10-ton DCC No 1692, built 1923; fifteen McLaren rollers were built for United Kingdom use, the last in 1924. Here, at Whitley Bridge, Yorkshire, in 1948, the owners were Osgoldcross Urban District Council. A manually-worked tar boiler is on the right.

(*Above*) Robey 'Lion' type DCC No 43693 with balanced piston valves, direct worm and quadrant steering and a graceful forecarriage cast in one with the smokebox. Shown working at Southwold, Suffolk, in 1948 when owned by East Anglian Roadstone & Transport Co Ltd. A total of four 'Lions' were made until 1934. (*Below*) Robey tri-tandem, one of three rebuilt from tandems for Wirksworth Quarries Ltd. The rear rolls are seen chain-driven together. The engine was DCC piston valve with ordinary Stephenson's link reverse which, with a small flywheel, gave a quick reverse. The boiler was Robey's stayless with round firebox.

Ruston & Hornsby (Lincoln) DCC No 143971, slide valves Class 'SCR', built c1926. This conventional design had a very good reputation for steaming and was made c1914–29. A Morrison sacrifier is shown here when owned by the West Riding of Yorkshire County Council (No 29), working at Whitley Bridge, Yorkshire, in 1948.

OPPOSITE PAGE (*Above*) Ruston & Hornsby 1937 and final design mainly for Far East use, made by Ruston & Hornsby and also sold by Aveling-Barford and Ransomes. Shown new c1950. This was a duplex cylinder engine with central piston valves. The roll head and smokebox were cast in one and the quadrant/worm gear steering was totally enclosed. (*Below*) In 1890, Thackray of Old Malton, Yorkshire, converted a Willsher 1878 traction (built by Fisken's of Leeds) to the roller seen here. The undermounted SC drove the crankshaft which passed through hollow studs on which the rear wheels revolved. The flywheel shown therefore revolved within and concentric with the wheel but at a different speed!

Wallis & Steevens 'Simplicity' type, No 7936, just under 3 tons with SC, one speed and inclined boiler with circular stayless firebox. Introduced in 1925 and intended for the Far East, it was not a selling success and only fifteen were made. Shown near Ambleside, Westmorland, in 1946 when owned by J. Chaplow & Sons.

Wallis & Steevens 'Advance' type No 7947, built 1927 and the final W&S design made from 1923 to 1940. Fitted with duplex cylinders having piston valves between them, disc cranks (no flywheel), quadrant steering and adjustable rear rolls for camber, this instant-reverse roller design remained in use to the end of commercial steam rolling.

INDEX TO PLATES

Sentinel tractor No 3237, frontis

TRACTION ENGINES

Allchin TE No 1285	9
Armitage & Ruston TE	23
Aveling & Porter RL	10
Bray TE	11
Brown & May SRL No 8742	10
Burrell TE No 1297	12
Burrell TE No 2836	14
Clayton & Shuttleworth crane TE	13
Cooper digger	13
Foden TE No 9052	15
Foster TE No 2827	16
Fowell TE No 99	17
Fowler Crane TE No 9713	16
Fowler TE No 3190	18
Garrett TE No 32890	19
Marshall TE	20
McLaren TE No 112	20
Ransomes TE No 8980	21
Robey TE No 29330	22
Ruston & Hornsby TE No 163688	23
Savage TE No 388	24
Wallis & Steevens TE No 7496	25

STEAM TRACTORS

Aveling & Porter tractor	28
Aveling & Porter tractor	29
Aveling & Porter tractor No 6898	29
Burrell tractor No 3851	30
Clayton & Shuttleworth tractor No 36921	32
Foden 'D' tractor No 14080	31
Foster tractor No 14378	32
Garrett 'Suffolk Punch' No 33180	33
Garrett tractor No 30988	34
Sentinel tractor No 567	35
Wallis & Steevens tractor No 2927	34

PLOUGHING ENGINES

Aveling & Porter PE No 8890	38
Fowler-Goode PE No 3938	39
Fowler SC PE No 2693	40
Fowler SC PE No 2050	40
Fowler DCC PE B4 No 8231	41
Fowler DCC PE K14 No 13452	42
Fowler DCC PE AA7 No 15257, second picture No 15257	43
Fowler SC PEs Nos 1071/2	44
Fowler DCC BB1 No 15154	45
Fowler DCC BB1 No 15409	46
McLaren DCC PE No 1552	46

SHOWMANS ROAD LOCOMOTIVES

Burrell SRL No 2804	49
Burrell scenic SRL No 4030	50
Burrell scenic SRL No 4021	50
Burrell SRL No 2780	54
Burrell S tractor No 3497	54
Fowler SRL B6 No 14424	51
Fowler S tractor No 14412	52
Fowler SRL No 15116	53
Wallis & Steevens SRL No 7125	55
Wallis & Steevens SRL No 2643	56

STEAM WAGONS

Allchin o/types	60
Atkinson u/types Nos 19/25	61
Aveling & Porter o/type No 8687	62
Burrell o/type	64
Clayton & Shuttleworth o/type No 48683	62
Clayton & Shuttleworth o/type No T1092	65
Foden o/type	66
Foden o/type	66
Foden u/type 'E' No 13180	67
Foden u/type 'O' No 14028	68
Garrett u/type No 35454	69
Hindley o/type	63
Hindley o/type (test)	63
Leyland u/type No F2/72/16822	69
Londonderry u/type No 38	70
Mann o/type	73
Savage u/type 'OA'	74
Sentinel u/type bus	71
Sentinel u/type 'DG8' No 8568	71
Sentinel u/type 'S−4' No 9227	72
Wallis & Steevens van	74
Yorkshire 'WG' No 2181	75

STEAM ROLLERS

Allchin roller No 1131R	78
Armstrong-Whitworth roller	79
Aveling & Porter roller No 12545	80
Aveling & Porter roller No 3798	81
Aveling & Porter convertible No 1262	82
Aveling & Porter convertible No 2527	83
Brown & May roller No 8130	83
Burrell roller No 4083	84
Clayton & Shuttleworth roller No 34380	85
Fowler roller No 15984	85
Garrett roller No 34084	86
Green roller No 1657	87
Marshall roller 'S' No 79260	88
McLaren roller No 1692	89
Millar-Marshall roller	87
Robey roller No 43693	90
Robey roller, tri-tandem	90
Ruston & Hornsby roller No 143971	91
Ruston & Hornsby roller (export)	92
Thackray roller	92
Wallis & Steevens roller No 7936	93
Wallis & Steevens roller No 7947	94